I0116407

# Interview—June 22, 2022

# Hrvoye Morić and Ken Coffman

Other books by Ken Coffman

*Steel Waters*
*Alligator Alley (with Mark Bothum)*
*Twisted Shadow (with Mark Bothum)*
*Glen Wilson's Bad Medicine*
*Toxic Shock Syndrome*
*Immortality, LLC*
*Hartz String Theory*
*Endangered Species*
*Fairhaven*
*Mesh (with Adina Pelle)*
*The Sandcastles of Irakkistan*
*Fianchetto*
*The Reluctant Queen (with Kristen Lolatte)*
*The Moon Maiden (with Kristen Lolatte)*
*Fiona and Minnie: The New Age (with Kristen Lolatte)*
*Buffoon*
*Real World FPGA Design with Verilog*

*Interview—June 22, 2022* ©2022 Ken Coffman and Hrvoje Morić All Rights Reserved

ISBN Print 978-1-949267-91-4

This book is sold subject to the condition that it shall not, by way of trade or otherwise, be lent, resold, hired out or otherwise circulated without the publisher's prior consent in any form of binding or cover other than that in which it is published and without a similar condition including this condition being imposed on the subsequent purchaser.

# STAIRWAY⹀PRESS

APACHE JUNCTION
**www.StairwayPress.com**
1000 West Apache Trail Suite 126
Apache Junction, AZ 85120

Hrvoje Morić is a Global Perspectives teacher at Nazarbayev Intellectual Schools in Kazakhstan. Previously, he worked as a Secondary Humanities teacher and Curriculum Coordinator at Prepa Tec and as an adjunct professor of International Relations at Tecnológico de Monterrey in México.

After obtaining a Bachelor of Arts in History and Secondary Education from Northeastern Illinois University, he earned a Master of International Relations degree from the Geneva School of Diplomacy in Switzerland. He has served as a volunteer with Peace Corps Mongolia and also worked as a staff assistant with the Mission of the Czech Republic to the United Nations.

Hrvoje has taught Business English, Economics, ESL, Global Studies and World History at the secondary level and numerous courses on International Relations at the undergraduate level. He is also a Gamification teacher trainer and heavily emphasizes technology in the classroom making use of applications such as Flipped Learning, Gamification, Reacting to the Past, Schoology and Skype in the Classroom. He is an Education Startup Digest newsletter curator and he produces the Geopolitics & Empire Podcast interviewing government officials and experts in international affairs.[1]

In 2017, together with 30 other American "citizen diplomats," he visited Russia for three weeks and met with academics, businessmen, citizens and politicians including former president Mikhail Gorbachev to discuss U.S.-Russia relations with the hope of bringing the Russian people's desire for peace to the attention of U.S. citizens and policy makers in Washington.

He hosts the worldwide Hrvoye Morić radio show on TNT Radio.[2]

He can be found on LinkedIn at:
https://www.linkedin.com/in/hrvojepmoric/

---

[1] http://www.geopoliticsandempire.com/
[2] https://tntradio.live/shows/the-hrvoje-moric-show/

Low-budget polymath Ken Coffman is a novelist, book publisher, Electrical Engineer, concert promoter and guitar shop owner (Lost Dutchman Music Company) who also wrote a nonfiction book (*Buffoon: One Man's Cheerful Interaction with the Harbingers of Global Warming Doom*) and a technical text (*Real World FPGA Design with Verilog*). His latest novel is #6 in the Glen Wilson series called *Immortality, LLC.*

Ken was born in Medford, Oregon. After living 50 years in the Seattle area, he currently lives in Sonoran Desert 35 miles east of Phoenix, Arizona.

He can be found on LinkedIn at:
https://www.linkedin.com/in/kencoffman/

Ken's Thermodynamics Library

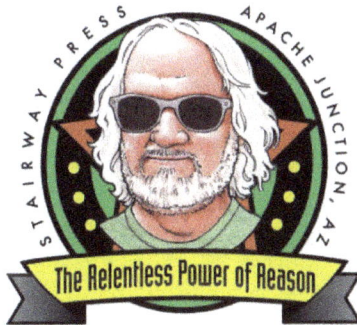

**Hrvoje Moric:** I'm joined by Ken Coffman, the owner of Stairway Press.[1] He's a Renaissance man with a fascinating background. He's been a berry picker, a cat food factory worker, dish washer, U.S. Air Force Sergeant, rock and roll bass player, concert promoter, author, publisher, electrical engineer and engineering manager.

**Ken Coffman**
Senior Field Applications Engineer -- Vicor New Space Initiative
Talks about #polymath, #bookpublishing, ##powerintegrity, and #progressiverockmusic
Greater Phoenix Area · Contact info
2,918 followers · 500+ connections

Open to    Add profile section    More

Vicor New Space Initiative

University of Washington

---

[1] www.StairwayPress.com

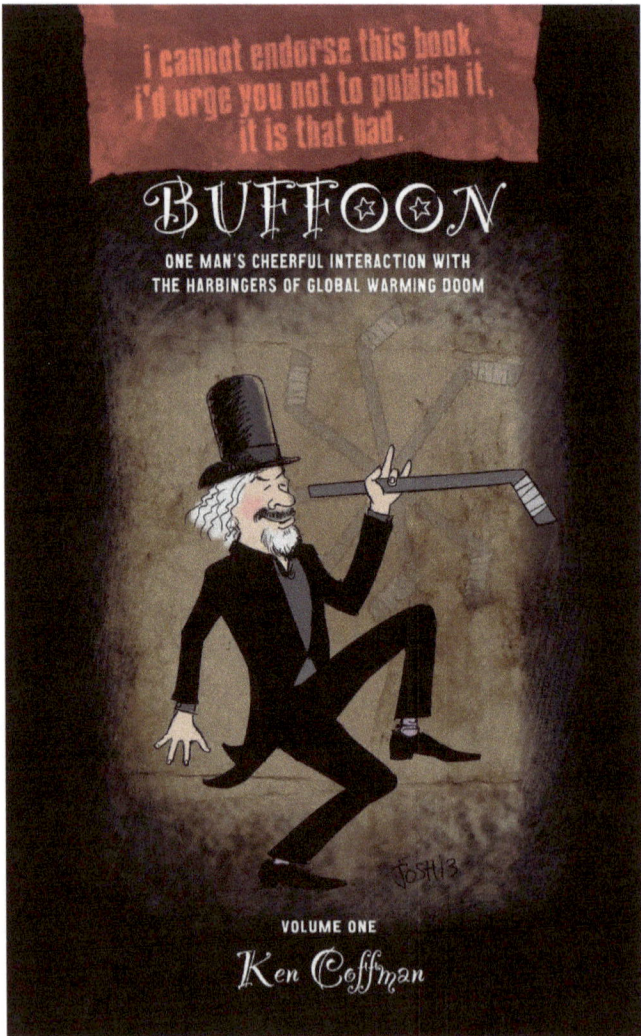

i cannot endorse this book.
i'd urge you not to publish it,
it is that bad.

# BUFFOON

ONE MAN'S CHEERFUL INTERACTION WITH
THE HARBINGERS OF GLOBAL WARMING DOOM

VOLUME ONE

Ken Coffman

He's written 12 novels, one nonfiction as well as a technical book. And his latest novel as far as I can tell is *The Sand Castles of Irakkistan*.[2] The website is stairwaypress.com. Go there and check out some of the books.

Hello, Ken. I'm jealous of your life experience.

**Ken Coffman:** You make it sound like I've accomplished a lot, but for a lot of those things you mentioned, I wasn't good at them.

**Hrvoje Moric:** I don't think that's as important as just having done them, but I guess that's subjective.

I know you write a lot and you've got Stairway Press and I really like these...I like TrineDay Press[3] and Clarity Press.[4] I'm a big fan of non-mainstream publishing houses.

Tell us a bit about Stairway Press and the work you've been doing lately.

---

[2] https://www.amazon.com/Sandcastles-Irakkistan-Ken-Coffman/dp/1941071058/

[3] https://www.trineday.com/

[4] https://www.claritypress.com/

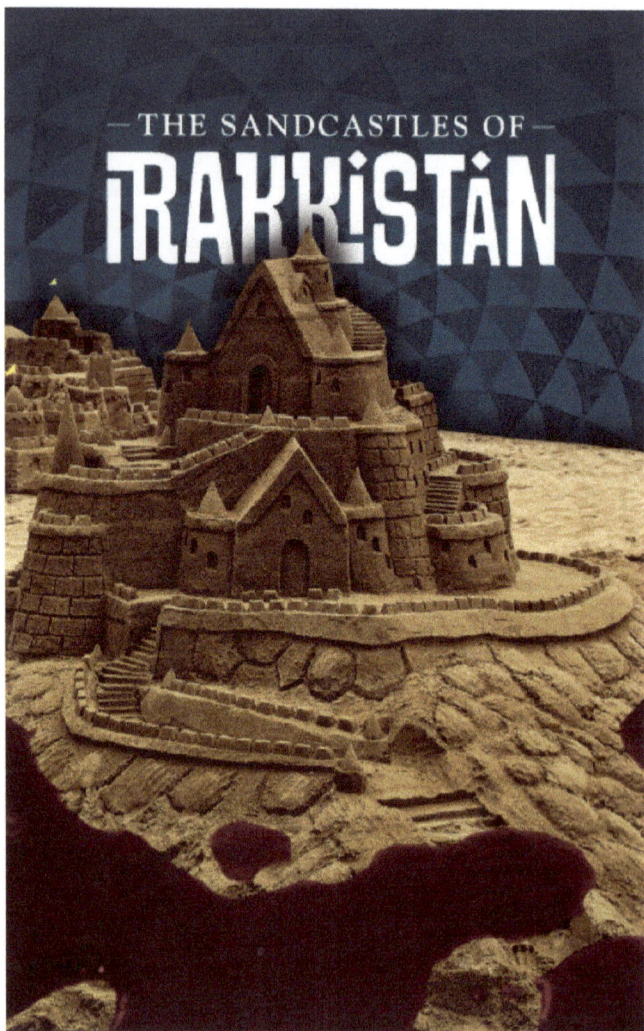

**Ken Coffman:** I'm 68 years old and I still have a day job which takes a lot of my time and attention. During the day, I work on satellite power supplies and radiation-tolerant electronics. That consumes a fair amount of my focus. I've been running Stairway Press…it's been pushing 15 years now.

Creating Stairway Press came from a passion for the written word. I'd been writing for a decade or so and got angry with collecting rejection slips. I literally have a stack one inch thick and I thought, "Well, I'll start my own publishing company." I quickly realized nobody would take me seriously unless I published other people's work. Almost immediately after that,

I realized I was happy selling other people's books. In fact, I was happier selling other people's books instead of my own.

If I sell you a copy of my book, I have an obligation to entertain, amuse or educate you—and I take that seriously. It's much easier to be objective about other people's work. I was able to select and work with people I chose and had skill and knowledge about things that interested me. I believe no one else in the world has published as many climate change titles as me. I'm not saying that we've sold more books than other publishers, but as far as having a variety of titles, I don't think anyone else has more. That started with my friends, John O'Sullivan, Tim Ball and the collection of authors called the Sky Dragon Slayers[5] back in 2011.

Over the years, I've been watching to find evidence that the ideas presented in that book, being 11, 12 years ago now, to find any errors. I'm not aware of a single one. An opinion is an opinion, right? You can have reasons for believing what you want, but for things presented as facts, they can be falsified.

If we present something as a fact and later evidence proves our thought was wrong, that's an objective criteria. And as I said, I've been watching very carefully over the years for evidence that the Sky Dragon Slayer ideas were wrong.

---

[5] https://www.amazon.com/Slaying-Sky-Dragon-Greenhouse-Theory/dp/0982773412/

Hrvoje Moric and Ken Coffman

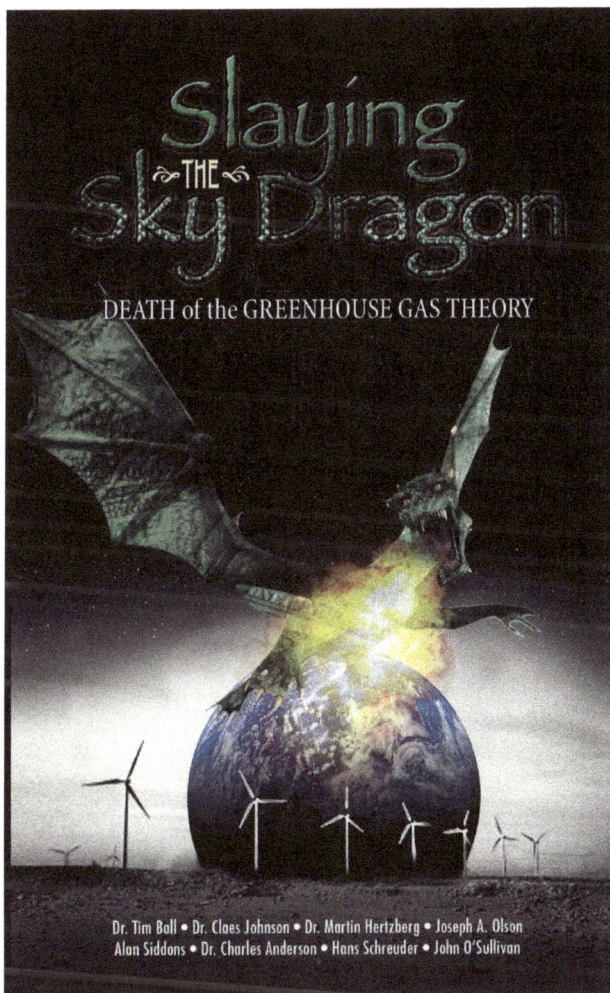

These errors do not exist. At this point, if you don't believe the climate hysteria is complete nonsense, then you're some mixture of corrupt or stupid—which is a bad combination.

**Hrvoje Moric:** I'm just realizing this...I feel like an idiot. I have the physical copy of *Slaying the Sky Dragon* right here behind me. I bought it 10 years ago or when it came out. I read it. I loved it. And I even communicated...I forget with which one of the scientists. The, "I love $CO_2$" guy. I don't know if you know which one. Was it Alan Siddons?

[The I Love $CO_2$ guy was Hans Schreuder, a co-author of *Slaying the Sky Dragon*. Hans is on the right in the photo below.]

**Ken Coffman:** Alan Siddons is brilliant.[6] It's interesting—

---

[6] *The problem all along, of course, is that people jump to conclusions. Sure, concentrated $CO_2$ exposed to infrared will get somewhat warmer than everyday air. But this only proves that everyday air (99.96% of which is nitrogen, oxygen and argon) is more transparent to IR and less apt to be heated that way. Air molecules, $CO_2$ included, initially acquire heat by contact with warmer surfaces. Via mutual collisions and convective transport, this heat gets spread around within an airmass.*

*To some slight degree, $CO_2$ also has the option of acquiring heat by radiative transfer. But, rather ironically, it cannot radiatively transfer this heat to the nitrogen, oxygen and argon molecules which surround it because, as said, they are largely infrared-transparent. As a result, an excited $CO_2$ molecule is obliged to share its heat just like the rest of them do, by bumping into other molecules. In short, there's nothing special about $CO_2$ in a real-world context. Outnumbered 2500 to 1, $CO_2$'s energy is lost in a busy buzz of collisions, its radiative properties wasted.*

*Moreover, any heated gas radiates infrared—and in this case 99.96% of the gas consists of molecules other than $CO_2$. Yet no one seriously imagines that back-radiation from 99.96% of the air has a role in raising the earth's surface temperature. Only when $CO_2$ comes up do we lose touch with reality.*

*Here's a succinct point: immersed in the vacuum of space, the earth has but one means of losing heat: radiation. And what does carbon dioxide do? It radiates. It's amazing that so few people have bothered to give this theory a second look.*

—Alan Siddons

he's so smart, but completely obscure. Nobody knows him. He's very difficult to contact, but he's one of the brightest, particularly in analyzing the arguments of the human-caused climate change so-called 'scientists.' Mentioning him is interesting to me. Some of these folks are heroes of mine and he's one of them. Tim Ball is another, these people are completely brilliant, completely honest and took huge risks in presenting their points of view when it was literally dangerous. It's not so bad now, but it used to be that your reputation and livelihood were at risk if you went against the climate gods...the established consensus of $CO_2$-based warming.

**Hrvoje Moric:** Hans sent me an "I love $CO_2$," sticker. I've got it here with me. I had no idea that was Stairway Press. I bought one of your books and I will look through the latest catalog. I saw interesting titles we can talk about in a moment.

*Slaying the Sky Dragon* is a great book. I used it in my courses I taught here in Mexico. I had a pair of students at a science fair here in Mexico try to debunk the human-caused warming narrative.

Hrvoje Moric and Ken Coffman

THE DELIBERATE
CORRUPTION
OF CLIMATE
SCIENCE

Tim Ball, PhD

I really have to give them props—these high school students. As you said, imagine pushing against the grain as high school students in academia. That takes courage because people look at you like you're crazy.

I've interviewed Lord Monckton, Rupert Darwall and recently Marc Morano and Tony Heller. I've been told you believe that science has been corrupted by an evil oligarchy and that Western academia is run by fools and crooks. And I totally agree. Maybe to get your inside, nuanced view on what's going on with the climate activists and green agenda. For me, it seems just like it's the old eugenics—Malthusianism dressed up in the color green—and it's just pretty much about control.

What are your thoughts?

**Ken Coffman:**   I agree. For young folks digging into this, we followed the same path initially with the Al Gore movie[7] and all of this kind of stuff, I was not critical about it at all and never even thought about it. It makes sense, right? We'll increase the concentration of $CO_2$ in our atmosphere and our average surface temperature will go up, right? That sounds compelling and scary. I envy the young people studying this, but I also dread the path they will follow. Once you figure out the human-caused global warming theory is not even close to accurate. Honestly, it's laughable to think human activities increase the Earth's average surface temperature.

---

[7] *An Inconvenient Truth*, 2006

Once you realize, first of all, how easily the average person is bamboozled and secondly, how corrupt people, like Michael Mann, are. For example, Dr. Mann, he's not a stupid person. He did his undergraduate work at Berkeley, a very tough school to succeed in.[8] He's a smart person. So, how has he been corrupted to believe this stuff? And does he *really* believe it? Or, is it just a way he found of making a career, getting paid really well and getting lots of self-serving publicity? I would love to lock him in a room and talk to him for a couple of days and try to figure out if he knows how wrong he is and if he's just taking an easy path to wealth, fortune and publicity, or if he really believes this stuff.

A scientist—someone who is objective and looks at data and evidence and carries it to a logical conclusion—there's no way that you can get there. It's embarrassingly stupid at this point to believe that increasing $CO_2$ in the atmosphere results in automatic heating. Think about it for a second. If you take a $CO_2$ molecule and heat it up, it will get more vigorous. Heat is a property of mechanical motion. You heat it up, it's going to move more. And naturally, you know this from your own personal experience, it will conduct its heat energy outwards, similar to thermals vultures drift on, similar to heat waves you

---

[8] Dr. Mann received his undergraduate degrees in Physics and Applied Math from the University of California at Berkeley, an M.S. degree in Physics from Yale University, and a Ph.D. in Geology & Geophysics from Yale University.

see emitting from a hot road. Your own direct experience leads you to knowing that adding $CO_2$ to the atmosphere results in additional cooling, if anything.

So, first of all, it doesn't do much, right? The radiative effect of atmospheric $CO_2$ is probably not measurable.

But considering the total overall effect, it has to be cooling because it's absorbing thermal energy and convecting outwards, taking that energy up with it. That's a cooling effect on the surface. It's absurd to conclude human activity increases the Earth's surface temperature. The idea is stupid.

We know there's an urban heat island effect and that cities are a couple of degrees warmer than the surrounding country side. That's a different thing because you're changing what's absorbed from the sun. And, apparently people want it to happen because they're living in cities. The cities are a little warmer. If they really hated additional heat they would live in the country instead.

**Hrvoje Moric:**  Yes, I agree. I read *Slaying the Sky Dragon* many years ago and I've seen the BBC documentary, *Great Global Warming Swindle*. And when you actually study this stuff, you realize everything they're telling you is junk.

Warming actually comes first and then $CO_2$ follows and then most of the $CO_2$ comes from the oceans and it's basically the sun driving everything. It's out of our hands.

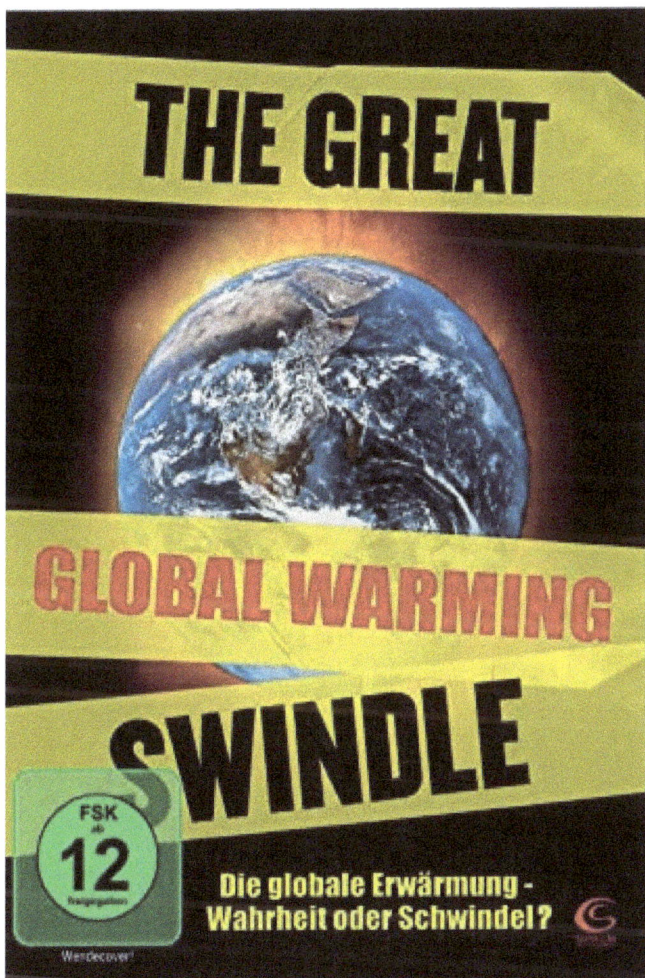

I taught a course called *Environment and International Relations* and I made a point...Piers Corbyn, the astrophysicist, brother of Jeremy Corbyn, he does the weather action. And supposedly farmers rely on his forecast, which he bases on the solar cycle and the lunar cycle.[9] [10]

I bought one of his forecasts to show my students. I bought it in January or February where he projected out into March and it sat there. And then to the day he said, "March 25th, you're going to have these crazy rains and floods in America, in the Midwest," and low and behold, MSNBC news, March 25th, was saying, "Historic flooding in Iowa," and then here and all over. And how did he do that? To me, that proved that his hypothesis is correct. If he could predict the weather like that, then that means it's the solar and lunar cycles creating all of this, not us. And yeah, that was amazing.

Let's take our first tequila break. The website is stairwaypress.com. Go there and get some books. We'll be right back on TNT Radio.

**Hrvoje Moric:** StairwayPress.com. Go there and get *Slaying the Sky Dragon*, as well as other books. You mentioned

---

[9] http://weatheraction.com/

[10] It's interesting that The Farmer's Almanac can accurately predict weather patterns. Their algorithm has nothing to do with atmospheric carbon dioxide.

See: https://www.farmersalmanac.com/predicting-weather

academia—I want to get your thoughts.

I worked as an adjunct professor teaching international relations and I found, whether it's a topic of climate change or other things like 9/11 or whatnot, you have official narratives that are not true according to people who seriously study these things.

I worked at a very globalist institution in Mexico. I mean, I think it's actually officially linked with the World Economic Forum. I'm not even joking. Tecnológico de Monterrey where I used to work is owned by a conglomerate called FEMSA, which is officially linked with the World Economic Forum.

They had Hillary Clinton and Bill Gates give speeches at the graduations, so they are definitely globalists.

But I find that most of the people in academia are either useful idiots—they've drank the Kool-Aid and they believe it—or you've got a second group that knows something's up, but they don't want to lose their career and the good money and prestige and that whole path. They'll be there for the rest of their life, so they don't want to go down the rabbit hole, even though they know something is up. And then I think there are some actors that are actually sinister and know the truth and either they're paid or psychopathic and pushing the globalist agenda.

What are your thoughts on the corruption in academia?

**Ken Coffman:** I agree with you. You mentioned the term psychopath. I think psychopaths tend to have a superior attitude. They enjoy the idea of getting publicity, being looked up to as a figure of authority and that kind of thing. Academia will attract that type of personality. You mentioned your young students and I said I admire them for having open minds and reevaluating the narratives being fed to them. But I also feel sorry for them because once you're been exposed to complete nonsense a lot of people accept without questioning, you lose confidence in your establishments, your universities, your government, big business and pharma. You'll lose a lot of stuff that might have been comforting.

There's definitely a dark side to pulling back the curtain and looking to see what's behind. It's like a five-year-old who figures out that Santa Claus can't possibly be real because of all the logical inconsistencies, then having to face a new reality that there're very few people you can trust.

The power of persuasion, the fact that it works on us as herd animals, really is scary. It's truly scary. We are animals, emotional animals. We don't tend to analyze. We tend to make decisions first and then invent reasons why the decision was logical. That's just human nature. So, anyone who studies what's been happening and recognizes the complete nonsense pressed upon us to believe, it's bound to make you lose hope for the positive potential of humanity.

**Hrvoje Moric:**   I think that's necessary. That's been a thing of mine to realize the nature of people. We have to understand and realize how bad it is. You have to go through that and then grasp all of this. Yeah, it's dark, but we then have the strength to face it, fight and not run away. It is depressing, but we have to be courageous to fight on all fronts.

What are your thoughts of what's behind this climate fraud? I mean, is it just money or power or what?

**Ken Coffman:** It's a combination of all that, and publicity, right? Publicity hounds want to their name in lights or in newspaper headlines. They want to be looked up to as an authority and it doesn't matter what the content is. If you can get this position of authority, you'll take it, however you can get it.

Think of the billions and billions and billions of dollars that have been spent on the green agenda. That's the bottom line. But we have convergences, harmonic convergences of power. Right now, it's social media, government and corporate interests.

The COVID-19 lockdowns did an excellent job of killing small businesses. So now, where will go for the stuff that you want to buy? Small brick and mortar businesses were decimated, killed. You have to go to the big guys. You have to search for your stuff with Google. You have to buy your stuff from Amazon. You have got to get your operating systems from Microsoft. The lockdowns were a huge boon for these guys.

Imagine that COVID-19 was bad for Twitter or Zoom's businesses. The publicity and the data presented would've been completely different. Maybe they didn't invent the pandemic crisis. They weren't smart enough to create this plan, but they were certainly smart enough to take advantage of it when it dropped on their lap.

**Hrvoje Moric:** Yeah. You mentioned the COVID-19, that was my next question because a lot of these things segue into the other. You had the pandemic lockdown and now they're segueing into climate lockdowns in France. They're saying it's so hot now, it's forbidden to have outdoor events. That's a climate lockdown basically. There's a title at Stairway Press called *COVID-1984: How the States of the World Destroyed Liberty*.

I recall way back in March of 2020, I had been interviewed by Spiro Skouras[11] who had a great, very popular channel, independent media, which got unfortunately deplatformed, and I started using that term COVID-1984 way back in March of 2020.

What are your thoughts on the past two plus years of this crazy biosecurity state? The response has been insane. It's like living in a science fiction dystopia, *Total Recall* and *Minority Report*, and it's just nuts.

It doesn't seem to be going away because now they're talking about monkeypox and cholera and polio. And

---

[11] https://truthcomestolight.com/author/skouras/

tomorrow Tedros [12] is meeting at the WHO to decide to declare monkeypox a DEFCON 5 and lock us down again. So, what are your thoughts of COVID-1984?

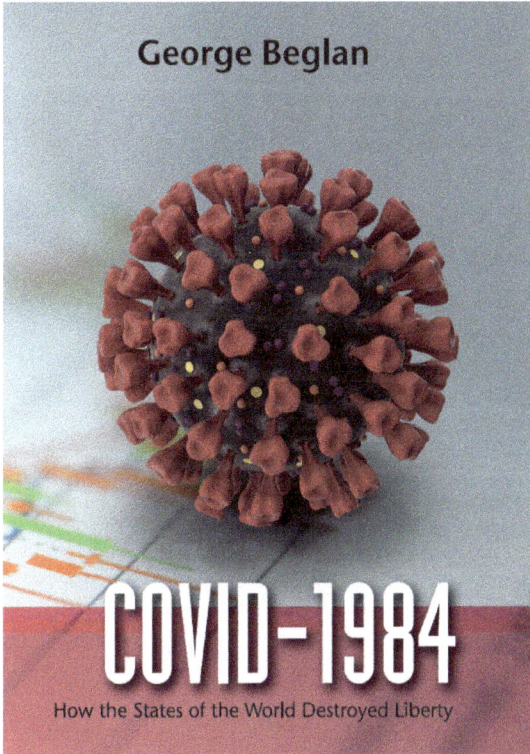

George Beglan

# COVID-1984

How the States of the World Destroyed Liberty

---

[12] Tedros Adhanom Ghebreyesus, Director General of the World Health Organization

**Ken Coffman:**    I like the meme that says, "Make Nineteen Eighty-Four fiction again." Other people have commented that the books, *Animal Farm* and *Nineteen Eighty-Four* seem to be used as guidelines or road maps. Boy, is that scary. Now, Orwell, Eric Blair, he was a smart guy. He reported on what he saw. So these things are basic human nature that occurs over and over again. It makes you lose confidence in your fellow man.

I remember talking to my friend Robert Ferrigno[13] one day and he said, "Look, if somebody put a gun to my head and said, convert to Catholicism or convert to Islam or convert to anything, what do I going to do? I'm going to say, 'yes, sir.' And I'm going to do it."

I was thinking about that. We would like to think we would stand by our principles and take the consequences. But the fact of the matter is most of us—and me included—would probably go along with it.

And that's what happens, right?

You'll be ostracized if you go against the central narrative. They have a lot of weapons at their disposal to destroy you. So it's safer to go along with the mainstream. We have families; we have people to take care of. It's safer to just go along. In the back of your mind, you're going, "I know this is complete nonsense," but the risk of going against it is just

---

[13] New York Times bestselling author of the *Sins of the Assassin* trilogy. http://www.robertferrigno.com/

not worth it. It is a Stalinistic type of environment we're living in. You would think we would learn those lessons and avoid it at all costs. But the ability to wield power, the propaganda power, over the masses is too compelling and works too well. You can't expect the elites to not use these tools. They're doing it. It's an amazing and a scary thing to observe.

**Hrvoje Moric:** And what about those people? I think there's that scene from *Nineteen Eighty-Four*, where they threaten to put Winston Smith's head in a cage so rats will eat his face.

My favorite dystopian novel is *We*, the original, by Yevgeny Zamyatin, that was produced in the 1920s. Huxley and Orwell riffed off of his original work. He's the original gangster though everyone cites Orwell.

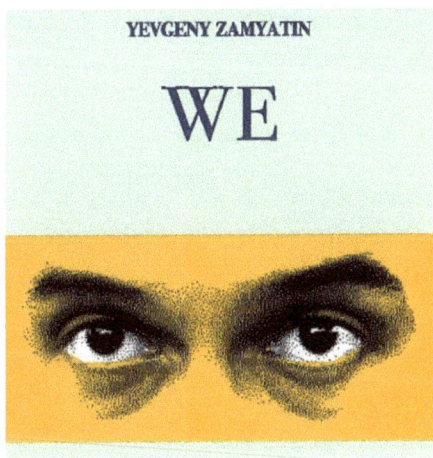

But I'm just one of those people that…I cannot say two plus two equals five. Let the rat eat my face. I'm not going to wear a mask. I just can't do it, come hell or high water. I think there were a number of those people. There's a local supermarket here, small market, fresh market near my home in Mexico. I didn't step foot in it for two years because the government required them to require us to wear a mask. And so I would find these other Mexican *tienditas*, corner shops, where I would buy my stuff there because they didn't really care if you wore a mask.

Any thoughts about pushing back against this, as you said, Stalinist sort of tyranny?

**Ken Coffman:** We need to fight on every battleground where we can engage in battle, but it's not easy. In Stalin's reign, there were people who resisted and spoke up and did different things against the state, but their fates were not good, right? I mean, typically they would find themselves in the Gulag. They would check in and they wouldn't check out. So I don't know. It's scary. I'm really afraid.

**Hrvoje Moric:** This is also related, another title at stairwaypress.com is, *Staying ON During the Great Reset* [14] which seems to be another aspect of all of this.

---

[14] https://www.amazon.com/Staying-During-Great-Reset-Patten/dp/1949267857/

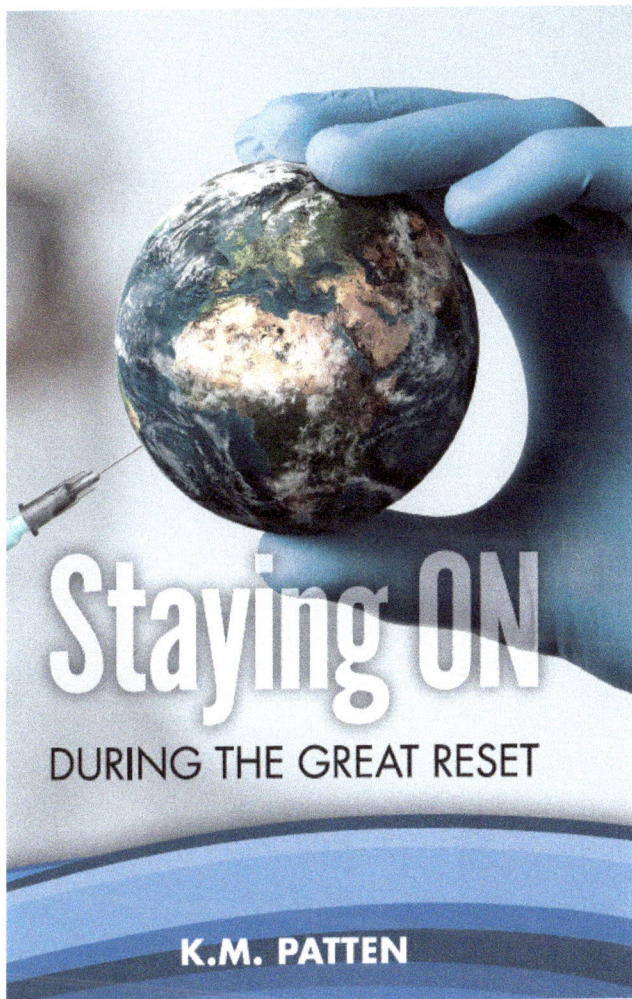

For me, the biggest thing is this push towards digitalization, the fourth industrial revolution. But we had the vaccine passports, which then morphed into this digital ID, digital passport and social credit system in the Western world, basically to the rest of the world outside of China.

And we saw...was it this week? Last week? In China? They got the Chinese people to start using digital QR code health passes, which I think originally the Nazis developed. And some of these folks in China, separately, there's a bank with financial problems. They froze people's accounts.[15] It's been frozen for a few months and a bunch of people...they don't have money. They need their money. They were going to go protest and the government turned their green health status to red and they could not travel on planes, trains, buses, and now we're seeing how—

**Ken Coffman:**   Yes, they couldn't gather in groups.

**Hrvoje Moric:**   Right. What are your thoughts on this other aspect of where they're trying to take us?

---

[15] https://www.reuters.com/world/china/china-bank-protest-stopped-by-health-codes-turning-red-depositors-say-2022-06-14/

# Check in now

**NHS**
Test and Trace

(!) Before entering, everyone must scan with the NHS
COVID-19 app or provide their name and contact details

**The Drapers Arms**

29-31 Cowgate, Peterborough PE1 1LZ

The NHS COVID-19 app is shown to help
stop the spread of COVID-19.

Protect your loved ones. Use the app.

**Ken Coffman:** It's horrifying enough to think about that happening in China, but think about that happening in Canada...the events of last year. The truckers peacefully protest and their bank accounts are frozen, they're being fired, they're being ostracized. This is not supposed to happen in Western countries, but it is, and it's accepted by a majority of people—or enough people.

The direction, vector and the intensity of this oppression is really something to observe.

**Hrvoje Moric:** Yeah. It's really crazy. Do you have any thoughts on with the war aspect, foreign policy? I haven't read the book, but you wrote on Irakkistan, which I assume is a novel touching on the theme of Iraq.

Way back in the day I had a girlfriend from Baghdad, Iraq and later on, it's funny, I studied graduate school in Geneva and one of my professors of international human rights law was Curtis Doebbler,[16] the Dutch American, who was Saddam Hussein's defense lawyer in 2003. He defended Saddam Hussein before he got hanged.

In any case, that was a war started for no reason that killed millions of innocent people. What are your thoughts on American foreign policy as well as the Ukraine situation, which has the potential to turn into World War III?

---

[16] http://doebbler.net/

**Ken Coffman:**   Which seems to be part of the point, right? In my book, *The Sandcastles of Irakkistan*, when I sat down to write it, my thought was, okay, what would it really take to resolve the tensions in the middle east? I'm just thinking things through. Basically, what happens in that book is rogue army soldiers are brutal and create a strong man who becomes a leader. They're able to take steps toward peace through brutality in the Middle East. That was what I was thinking. Some of these problems are not solvable. People are entrenched in their beliefs, in their history and culture and they're not going to change. You have these fault lines.

What really bothers me more are the fault lines that are intentionally emphasized. Some of these disputes, whether abortion, or…pick the hot topic of gender nonsense and all this stuff. Because they can't be solved, they're fault lines that can be exploited. They're exploited by our side and they're exploited by the other side. I was just reading a book about Russian covert activity in the United States.[17]

They basically look for any chance to cause trouble—and there're lots of ways to cause trouble. They just made sure to do everything they could to put gasoline on the flames. We're ripe for it.

---

[17] A trilogy by Tom Rob Smith that starts with *Child 44*. https://www.amazon.com/Child-44-Tom-Rob-Smith/dp/0446402389/

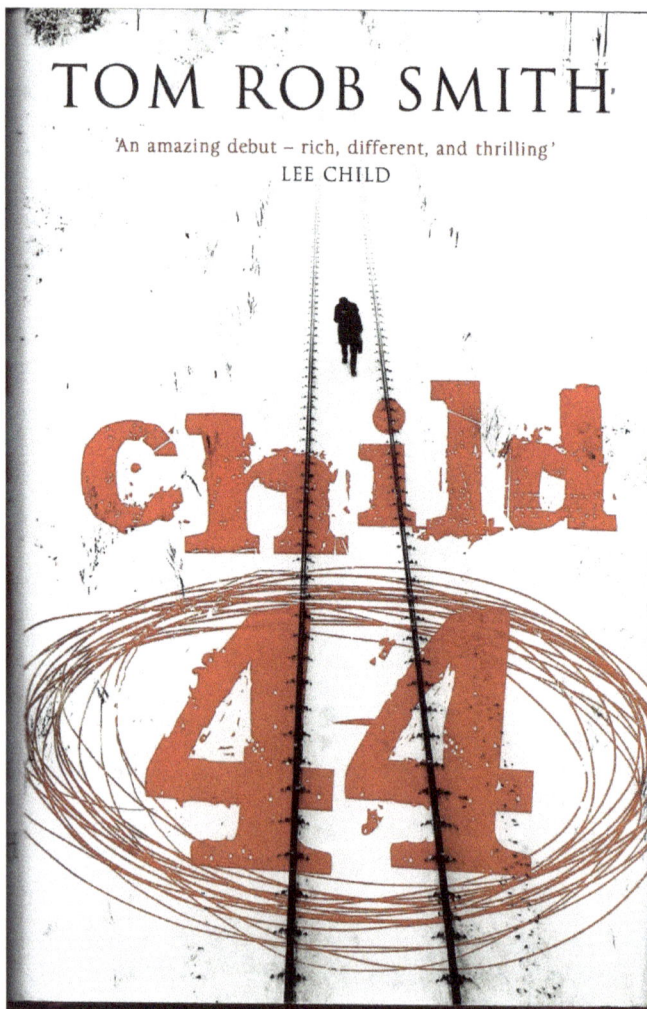

# TOM ROB SMITH

'An amazing debut – rich, different, and thrilling'
LEE CHILD

child

44

I don't want to paint a truly dismal picture because I believe it's still possible to tunnel underneath all the nonsense and have a great life, but generally speaking, the way humanity acts, there's no positive outlook.

It's human nature to press a system until it breaks. We exploit systems and abuse them, push, push, push. We never evolve toward anything good. What happens is we push and push and push, the system breaks down, and then at that point you've got a generation of chaos, which is worse. But after that generation of chaos, things can get better again.

It just doesn't paint a very positive picture for humans to be able to think their way out of our problems. The tools and the power are in the hands of psychopaths, as you said, and they're not going to allow a positive outcome for regular people.

For example, I believe we could feed everybody on the planet. We could make sure they have a roof over their heads. We could do that for everyone who exist today, but we're not going to because it doesn't serve the purposes of the elite. It doesn't serve the purposes of the rich. Too many people are, like Bill Gates says, useless eaters. We have to put up with them until we can kill them, or do whatever it is they're going to do, wipe us out with some factory-generated disease.

I don't know.

I'm actually a very cheerful person, but when thinking about the exploitation, the nonsense and the propaganda and the power that's held by the psychopaths, it's hard to see how we get out of this gracefully. I don't think it will happen.

**Hrvoje Moric:** Yeah. I kind of have that same view. Daniel Horowitz had a tweet today, he said, "The new October surprise is unleashing another gain-of-function pathogen on us. These people are doing worse than any of the bio terror we envisioned coming from Al-Qaeda after 9/11."

We're going to our second tequila break. While we're there, you can check out Stairway Press website.

We are back on the Hrvoje Moric show on tntradio.live, talking to Ken Coffman of StairwayPress.com.

It was funny, I didn't even know I had one of his books that I bought a decade ago. *Slaying the Sky Dragon.* Excellent book. And to talk a bit more about what you were mentioning before the break, whether people are optimistic, pessimistic or even realistic.

I'm a conservative, I believe in the Bible and the prophecies and all of that. I kind of feel like at some point, that stuff is inevitable.

So in the material plane, we are headed in a progressively worse direction over time. When the end is, I can't tell you, but I feel like we're on this long slope into tyranny and dystopia and things are bad.

But meanwhile, what can we do?

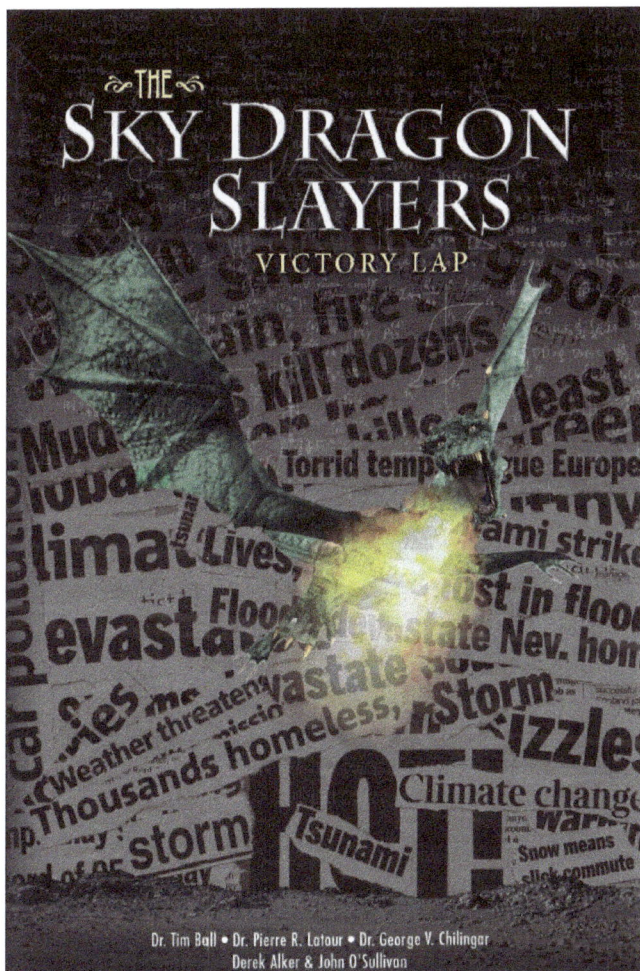

There're people building decentralized structures, parallel structures like in the old Soviet days. Under tyrannical regimes, people were building parallel economies, parallel societies, parallel structures. I see a lot of people doing that today. I meet a lot of people fleeing Canada, fleeing the U.S. and Europe and the Anglosphere and coming to Mexico, or hear about them leaving the cities and moving to rural areas, buying farms, raising chickens, planting food.

What are your thoughts on this sort of thing?

**Ken Coffman:** I think that's exactly correct. The urban centers, the urban environments represent a huge amount of risk due to the complex logistics and extreme effort it takes to keep these cities going.

You have a lot of people who aren't super productive. What I mean is: they're not super useful. They might be good at computer programming or this, that, and the other thing, but they're not directly contributing to manufacturing, growing things or adding value to materials and turning them into something else.

So, I do think the move to being more rural and escaping is exactly a smart thing to do. It's not hard to imagine completely terrible things happening in San Francisco, in all of the big cities, if the stress gets higher and ratchets up and ratchets up and ratchets up until the system finally breaks down.

**Hrvoje Moric:**   Yeah. These are complex systems, the cities. I wonder how long they can last. I live in the second biggest city in Mexico and people like Jim Rickards[18] or others who talk about complexity theory, systems theory, at some point, it takes more energy required to put into the system than you get out and then you get collapse. So I wonder how long this stuff can go on.

I want to ask your thoughts about the economy as well, as you see it. A lot of people are talking about a crash, a historic economic crash. It certainly seems like we're headed for something really, really bad that none of us in our lifetime have ever seen or experienced. Inflation…every day, it's like 10% inflation in all these different countries, all over the world.

Sri Lanka just collapsed.

I can't even keep up with all this crazy news. All these new records. All 50 states in the U.S. have $4 a gallon gas, which was a record, and now we're moving on to $5. In California, some people pay $9 or $10 a gallon. We're seeing things that we've never seen in our lifetime. And then there's talk of hyper inflation and a second civil war in America. What are your thoughts?

**Ken Coffman:**   Imagine a city, like Mexico City, where half

---

[18] https://hiddenforces.io/podcasts/jim-jrickards-predicting-the-next-financial-crisis/

the people, half the population, are genuinely hungry—unable to get food. Follow that line of thought with all the major cities and you can see we have huge problems.

People take for granted the miracle of development over the last 150 years that's led to being able to sustain 8 billion people on the planet. It's a miracle, but the problem is that the people that made that happen, the engineers, the workers, the creators and all that kind of thing...that's not where you get rewarded these days. Social media and TikTok videos are what get you noticed. But, if you're hungry and need to find food, your TikTok video will not do you much good.

**Hrvoje Moric:**   There was a survey not long ago where they asked Americans and Chinese children what you'd like to be. The poll said that by far, American kids wanted to be YouTubers and the Chinese wanted to be astronauts—something more productive than being a YouTuber.

I can tell you, I never even wanted to be a YouTuber or a podcaster. I just fell into it. And I would say, don't do it, especially now in the environment that we're in of Orwellian censorship.

Are there any other thoughts pressing on your mind related to what we've been talking about or not, or perhaps related to some of the titles at Stairway Press?

**Ken Coffman:**   As you mentioned, the things that give me hope are people abandoning the system. I think there is a way to tunnel underneath all the nonsense and live a great life, but you'll have to keep your eyes open and might have to move a time or two to avoid serious problems.

But, if you're smart and wise, regardless of how bad things get, you'll still be able to have a great life. The home-schooling movement, there's a lot of potential for that. There are a certain percentage of people who realize the established, conventional wisdom and conventional education, government education, government medicine, and all this kind of stuff, are basically poisoning you and that you need to avoid all that.

I want to mention a book I read recently. This is a non sequitur, but I was curious about Kary Mullis who won the Nobel Prize in Chemistry for the PCR test. I got his autobiography and was completely fascinated by it because he was such an interesting character, an iconoclast and free thinker. People looking for something cheerful, interesting and fun, I highly recommend they get Kary Mullis's book.[19]

He talks a little bit about the PCR and there wasn't a lot that blew my mind about that science, but the facts of his life, the way he lived and the way he thought, was very uplifting and very interesting. It's called *Dancing Naked in the Mind Field*. So, as ray of hope, check that book out. I think it would make you happy.

---

[19] https://www.amazon.com/Dancing-Naked-Mind-Field-Mullis/dp/0679774009/

DANCING NAKED
IN THE MIND FIELD

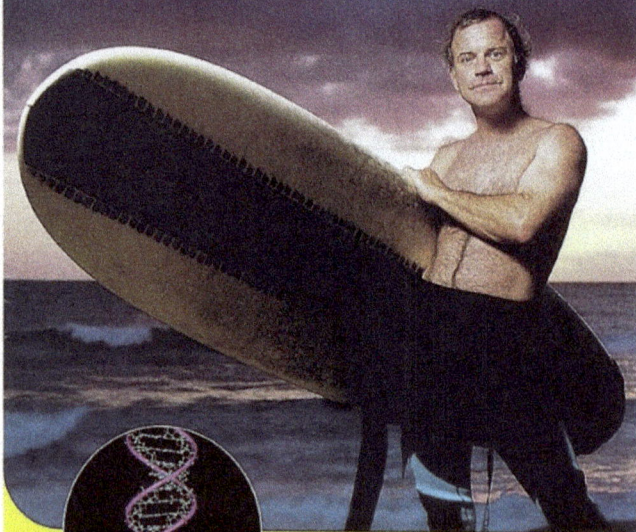

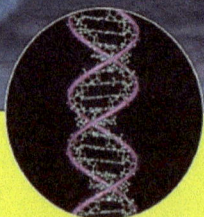

WINNER OF THE NOBEL PRIZE IN CHEMISTRY
KARY MULLIS

"Kary Mullis, perhaps the weirdest human ever to win the Nobel Prize in Chemistry, [has written] a chatty, rambling, funny, iconoclastic tour through the wonderland that is [his] mind."
—THE WASHINGTON POST

**Hrvoje Moric:** I haven't had the time, but I want to delve deeper into his work. It's strange that he passed in 2019, right before all of this COVID-19 stuff.

**Ken Coffman:** It would've been fascinating to see his take on what's happening and how the PCR test has been completely abused. If we get a chance to talk again, I would like to talk about post-modernism and why it is that young people think the way they do, why they don't believe in an objective reality.

I feel like I've landed on the reason and I'd love to talk to you in more depth about that topic and why young people think the way they do.

**Hrvoje Moric:** Definitely. That leads to the whole transgender thing. And it's all related: post-truth, post-modern, where I feel like some of these powers, they want to destroy the foundations of objectivity...of objective truth and science, things that exist. And once they destroy that, they can make us believe anything they want.

And I think transhumanism and transgenderism are a result of all of this.

**Ken Coffman:** They're doing it, right? They're pulling it off.

**Hrvoje Moric:** We're watching it in real time. Where are the best places to find you online or best websites to visit?

**Ken Coffman:** I'm active on LinkedIn and on Facebook for some reason, but mainly, just buy Stairway Press books or send us an email to ask us what we've been reading because I can point you toward books that will bring a little cheer to your life.

We need a little bit of entertainment and hope.

**Hrvoje Moric:** We do. All right. We're out of time. Stairwaypress.com. Check it out.

Thank you, Ken. Next time, hopefully, we'll talk about post modernism…

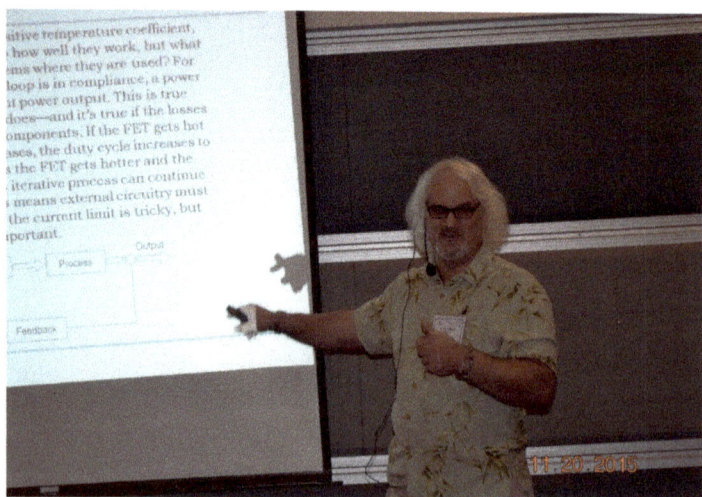

www.ingramcontent.com/pod-product-compliance
Lightning Source LLC
Chambersburg PA
CBHW051431270326
41933CB00022B/3486